Georg Gemünder

Georg Gemünder's Progress in Violin Making

With Interesting Facts Concerning the Art and Its Critics in General

Georg Gemünder

Georg Gemünder's Progress in Violin Making
With Interesting Facts Concerning the Art and Its Critics in General

ISBN/EAN: 9783743404304

Manufactured in Europe, USA, Canada, Australia, Japa

Cover: Foto ©berggeist007 / pixelio.de

Manufactured and distributed by brebook publishing software (www.brebook.com)

Georg Gemünder

Georg Gemünder's Progress in Violin Making

BIOGRAPHY

OF

GEORGE GEMÜNDER.

George Gemünder was born at Ingelfingen, in the kingdom of Wurtemburg, on the 13th of April, 1816.

His father was a maker of bow instruments, and it was, therefore, from Gemünder's earliest youth that he devoted himself to the same art and the studies connected with it.

When he left school, it was suggested to his father that George should become a schoolmaster, as he at the time wrote the finest hand and executed the best designs of any among his classmates. His father was not averse to this proposal and decided to carry it out. George was, accordingly, directed to prepare for the seminary. The plan was not, however, in accordance with his own tastes or inclinations, and he followed it for a period of but three weeks, only to abandon it finally and forever, to take up that employment which accorded with his natural gift and gave scope for the development of his genius.

After his father's death, which occurred when George was in his nineteenth year, he went abroad, and worked variously at Pesth, Presburg, Vienna and Munich. Fortune smiled upon him, and more than once an opportunity was presented of establishing a business; but nothing that promised simply commonplace results and a commonplace life could attract his eye, since his mind, aspiring to improvement in his art, was constantly impelling him toward that celebrated manufacturer of violins, Vuillaume, at Paris. He plainly saw that in Germany he could not reach in the art that degree of accomplishment for which he strove, and, therefore, he resolved to find, if possible, at Strasburg, such a position as he had had at Munich. Through the mediation of a friend he obtained a call to go to a manufacturer of musical instruments at Strasburg; but upon his arrival he was astonished to learn that the man was a maker of brass instruments! Here was a dilemma. Disappointed in his effort to find employment, winter at the door and far away from home, what could he do? The manufacturer, whose name was Roth, perceiving his perturbation, was kind enough to ask Gemünder to remain in his house until he should have succeeded in finding such a position as he desired. Gemünder accepted the profered kindness, and after the lapse of six weeks he formed the acquaintance of a gentleman with whom he afterward became intimate, and who promised to write for Gemünder a letter of recommendation and send it to Vuillaume at Paris. Meanwhile Gemünder remained in Strasburg. One day, while taking a walk in the park called "Die Englishen Anlagen," he seated himself on a bench and shortly fell asleep. In his sleep he heard a voice which seemed to say: "Don't give way; within three days your situation will change!" The voice proved prophetic, for on the third day after the dream his friend came to him with a letter from Vuillaume, which contained the agreeable intelligence that Gemünder should go to Paris. The invitation was promptly accepted and Gemünder immediately started on his journey. When he arrived at Vuillaume's another difficulty was encountered, for Vuillaume had mistakenly supposed that Gemünder spoke French. By mere good fortune it happened at the time of Gemünder's arrival that a German professor was giving music lessons to Vuillaume's twin daughters, who in the capacity of interpreter informed Gemünder that M. Vuillaume was sorry to have induced him to come to Paris, because it would be impossible to get along in his house without French. Vuillaume kindly offered to pay Gemünder's traveling expenses from Paris back to Strasburg, but said, however, that should the latter be satisfied with nominal wages at first, he would give him thirty sous a day until he should have learned enough of the language to be able to get along. Gemünder accepted the proposition, which greatly astonished Vuillaume because he had not supposed that Gemünder would be contented with such small wages! Then he showed him a violin and violoncello as models of his manufacture, and asked him if he could make instruments like those. The answer being in the affirmative,

Vuillaume smiled, for he was sure it could not be done. On the following day he provided Gemünder with materials for making a new violin, in order to see what he could do. He soon perceived that Gemünder possessed more theoretical than practical knowledge. When the violin was finished, he made him understand that their way of working was different, and he desired to have his own methods adopted. Gemünder did his best, and being a good designer, he soon acquired a knowledge of the different characters of the propagated Italian school in regard to the construction of violins.

After the lapse of three months Gemünder's wages were increased ten sous a day, and although he now saw his most heartfelt desire fulfilled, namely, to work in Vuillaume's manufactory, yet he did not find it possible to stay there permanently, because his fellow-workmen, who had observed the kindness with which their employer had treated his new workman, became filled with feelings of jealousy, and resolved to harrass him and compel him if possible to leave. So thoroughly did they succeed in embittering his life, that Gemünder finally resolved to leave Vuillaume and go to America, and with this firmly fixed in his mind he began his preparations secretly to carry out his plan.

When everything was ready, he went to Vuillaume to make known his intention and to explain to him the cause of his leaving. The latter, astonished at this intelligence, declared that Gemünder should not leave his house at all, and assured him that he would not meet with further unkindness from his fellow-workmen, even if all should be dismissed, although some of them had already been in his manufactory for many years. He further assured Gemünder that should he not desire to remain Paris, he would establish him in a business similar to his own, either in Germany or elsewhere, but he dissuaded him from going to America, for the reason that the art of violin making was not sufficiently understood there at that time. This kindness and benevolence upon the part of his employer so touched his heart that he was constrained to remain, and he began to construct new violins, in some of which he imitated the Italian character thoroughly, and also to repair injured violins.

One day Vuillaume handed Gemünder a violin, with the remark that he wished him to do his best work in repairing it, for a gentleman from Russia had sent it. Vuillaume especially called Gemünder's attention to a certain place in the back which was to be repaired, which was almost invisible, and he gave Gemünder a magnifying glass for his assistance, but Gemünder returned it, saying that he could do better with his naked eyes, and when finished Vuillaume might examine it with the glass. When completed, the work proved to be all that Vuillaume had wished, and satisfied the owner of the instrument so thoroughly that in his ecstasy of delight he presented Vuillaume, in addition to the payment for his work, with a costly Russian morning gown.

On the return of Ole Bull from America, in 1845, that distinguished performer brought his wonderful "Caspar da Salo" violin to Vuillaume to be repaired, and requested the latter to do the work himself, as it was something about which he was very particular; but Vuillaume answered that he had a German in his workshop who could do it better than he. Impelled by curiosity to become acquainted with this German, he asked to be shown to the place. After some conversation, Gemünder undertook the repairing of the violin and completed it in as masterly a manner as he did in the case of the Russian gentleman.

After an interval of three years, while Gemünder was still working at Vuillaume's, the latter showed him a violin and asked his opinion about it. Gemünder, having examined it, replied that it was made by some one who had no school! "I expected to hear this," returned Vuillaume, "and now let me tell you, that this violin is the very same that I engaged you to make when you came to me. I show it only that you may recognize what you are *now* and what you were *then !*" Gemünder was not only surprised, but amazed, and would hardly have believed it possible. This incident is only mentioned to show that as long as the eye has not been fully cultivated, those who fancy themselves to be artists are not such, and in reality they cannot distinguish right from wrong. Gemünder has often experienced this in America. He knows no other violin maker who deserves to be compared with Vuillaume in this respect, for he correctly understood the character of the outline and form as well as the interior structure of the different Italian instruments.

Towards the end of 1847, when Gemünder had been four years at Vuillaume's, his two brothers, who were in America, invited him to go there, as the interest in and taste for music was improving and they intended to give concerts. Gemünder therefore determined to accept this invitation and left Paris. He arrived in November, at Springfield, Mass., and, meeting his brothers, arrangements for concerts were made with an agent, who engaged several other artists to make up the company. The instrumental quartet consisted of a clarinet, violin, flute and bass guitar. This music made quite a sensation, and the houses were always crowded, yet the Gemünder brothers did not receive anything from the proceeds. They soon comprehended that they had had too much confidence in their agent, and after the lapse of a week they gave up the speculation.

For George Gemünder, who had then very little knowledge of the English language, which fact increased the difficulty of his position, there remained no other choice but to settle as a violin maker. He borrowed from a friend twenty-five dollars, and with this money he set out for Boston, Mass., and established himself there. The violins which he made he sold at fifty dollars each, and made repairs at low prices.

In 1851, when the first exhibition of London took place, Gemünder sent a quartet of bow instruments, in imitation of Stradivarius, and one violin according to Joseph Guarnerius, and another according to Nicholas Amati.

As his business in Boston did not prove sufficiently lucrative, Gemünder left the city after eighteen months, without waiting for news of the result of the exhibition, and established business in New York. Later he learned that his instruments had received the first premium at the exhibition.

When, in the following year, 1852, Gemünder received his instruments back from the exhibition, he learned that Ole Bull was in New York again, and, as he had formed his acquaintance in Paris, he paid him a visit and gave information that he had established himself in New York, and also that he had obtained the first premium at the London exhibition. Ole Bull was highly astonished at this news, as he said "Vuillaume is the best violin maker, and I have on one of my violins the best specimen of his workmanship as a repairer." He thereupon showed Gemünder his "Caspar da Salo." "Here," he said, "look at it, find the place where the repair was made." But

Gemünder replied: "Sir, have you entirely forgotten that when you went with your violin to Vuillaume, he made you acquainted with a German in his studio, whom he directed to repair this 'Caspar da Salo' violin, and that this German was myself?" Upon hearing this a light seemed to break upon his mind, and he exclaimed, "Yes, yes, I do remember. Now you shall become in America what Vuillaume is in Europe."

Meanwhile the advantages which might have been derived from the London exhibition were lost, in consequence of Gemünder's removal from Boston and establishing business at New York. Spohr, Thalberg, Vieuxtemps and many more of such authorities, examined his violins in the exhibition and were much surprised at the excellent qualities of the instruments. Spohr observed: "These are the first new violins that I ever saw, tried and liked!" When they were played upon by him and others, they attracted hundreds of admirers and would have been sold at high prices had Gemünder not failed to make arrangements to dispose of them.

The results obtained at Paris and Vienna were similar, his instruments attracting much attention in each exhibition. In the Vienna Exposition, held in 1873, Gemünder gained the greatest triumph that was ever obtained by any violin maker. The "Kaiser" violin sent by Gemünder in response to an offer of a prize for the best imitation, was declared by the professional judges to be a renewed original; a genuine Guarnerius not only in regard to its outer appearance and character, but also as to its wonderful quality of tone and ease with which the tones come. To find these qualities in a new violin was beyond all expectation, since it had hitherto been taken for granted that such a result could not be obtained, because that object had been the unsuccessful study of different makers for hundreds of years. This proves, therefore, to the musical world, that Gemünder has solved that problem which has generally been considered impossible. In spite of all this, however, Gemünder had learned by painful experience that the prejudice existing among most of the violinists was not to be wiped out. These people are incapable of judging reasonably, and it is easier for them to say that Gemünder makes his new violins of wood prepared by a chemical process, or that it has not yet been proven that his violins have kept their good quality for an extended period of time, notwithstanding that Gemünder has been constructing violins in America since 1847, and that nobody can prove that any violin of his making has lost its quality of tone. On the contrary, they have invariably proved good. Gemünder, however, confesses that a few of his first made violins in America do not equal those of his present construction in regard to tone and varnish. The cause of it was that Gemünder being unacquainted with the woods of the new country, was not so successful at first in the choice of wood for his violins, and naturally would not be until his experience had improved. The prejudice above referred to would, however, be likely to exist for another century, could Gemünder live for that length of time among those people, the most of whom would persevere in their opinions.

The impracticability of the theory of using chemically prepared wood for violins is sufficiently understood at the present time to render it useless to pursue the discussion in these pages. Gemünder has informed himself as to the degree of success attained in the use of the different chemical preparations of wood, as well as those prepared with borax, by which, the inventor asserts, the wood becomes richer in tone and lasts longer than that which is left in its natural state. Yet, without opposing the inventor, Gemünder follows the principle of the old Italian violin makers, because their productions have been in use to this day; therefore the material left in its natural state has proved good and has satisfied the musical world for these three hundred years. He has indeed succeeded in constructing new violins of material in its natural state, which produce not only an extraordinary power of tone, but also a strikingly equal quality of tone, and the quality of easy speaking, and the outward appearance of the old violins has been so faithfully imitated that he who has not been told the fact, will take them for genuine instruments made by Stradivarius, Guarnerius, Maggini, Amati, and others.

It is therefore assuming not too much to say that George Gemünder has surpassed in this art all the violin makers of the present and past times; for where the Italian masters ended with their knowledge, George Gemünder commenced and improved, which fact can be proved to the satisfaction of every critic ; for George Gemünder has not only gained the same results as those achieved by Stradivarius and others, but he has sketched a better acoustic principle for producing tone. It is for this reason that August Wilhelmj, the great violinist, calls George Gemünder the greatest violin maker of all times, for Wilhelmj had learned by ample trial of the instruments made by George Gemünder that they were incontestably all that the latter claimed for them. Wilhelmj admired Gemünder's " Kaiser " violin at the Vienna Exhibition, as it was the only violin of importance which attracted his attention, and this aroused within him the desire to become personally acquainted with its maker. By means of his renown as the great violin virtuoso, an engagement was offered him to go to America, which he accepted, and thus his wish was fulfilled. On the day after his arrival in New York, Wilhelmj went to see Gemünder at Astoria, and from that time has been Gemünder's friend and admirer.

Wilhelmj and other artists have expressed astonishment that a man of George Gemünder's capabilities in this art was to be found in America. Although he enjoys the highest renown in his art, yet he lives in a country in which the appreciation of that art is still in its development; for the number of amateurs such as are found in Europe, who spend enormous sums in instruments, is very small here. The fact is that George Gemünder lives here at too early a period, for his productions are a continuation of those which the great Italian masters brought forth. Taking into consideration all the foregoing circumstances it is fair to suppose that George Gemünder has had to contend with extraordinary difficulties during this long time. For ignorance and arrogance can do much damage, in this respect, not only to the artist, but also to the amateur, as these often times place their confidence in those musicians who who have no knowledge of violins, and who can only mislead them.

APPENDIX.

GEORGE GEMUNDER'S OBSERVATIONS IN REGARD TO VIEWS WHICH THE MOST OF VIOLINISTS AND AMATEURS HAD OF THE TONE OF OLD AND NEW VIOLINS—HOW THEY IGNORED THE NEW INSTRUMENTS, AND HOW THEY WERE DECEIVED AND SURPRISED IN THEIR PREJUDICE CONCERNING THEM.

Gemünder had learned that the knowledge of arrogant violinists and amateurs in regard to tone did not rest on any correct basis, and that their prejudice rested on a tradition arising from the decline of the manufacture of violins since the death of the celebrated Italian makers. All attempts of late years to make good violins having failed, an aversion to new violins has been gradually spreading, so that the most of people at the present time do not believe it possible for violins to be both new and good. Firstly, because it has been found that new violins have not been constructed so as to possess the tone of old Italian instruments; and secondly, that those made of chemically prepared wood did not stand proof for a great length of time. Many musicians and amateurs have in consequence of this prevailing prejudice gone to an extreme and disregarded new violins, no matter what tone they might have. To this class of people belonged especially the violinist Wieniawski, who had an opportunity to play on one of the best violins made by Gemünder, which opportunity he ignored, because the violin looked new. Instruments imitated by Gemünder were placed before him as genuine violins, and he admired them. Ole Bull was equally surprised when an imitation according to Stradivarius was handed to him in Columbus, Ohio, and he declared it to be a genuine original.

When Vieuxtemps gave concerts in America for the first time, and went to see his friend Vieweg, Professor of music in Savannah, Ga., the Professor showed him his Stradivarius violin. Vieuxtemps, catching sight of it, said: "If he had not been quite sure that his violin was at home, he would think it was his own." But when his friend told him it was a Gemünder violin, he was astonished and observed: "The d***l knows how Gemünder can bring such a tone in new violins!"

At about the same time a violinist came from Germany and visited Gemünder to hear his violins, because Spohr had praised him so much; but at the same time he doubted that new violins could sound like those of the old Italian masters. Gemünder first showed him some having the appearance of being new; the violinist played upon them and then uttered: "They are as I thought; they have not that sweet, melting tone of the Italian instruments." Hereupon he asked Gemünder if he had no Italian violins, in order to show the difference. Gemünder then opened another box, and showed him an imitation of Amati for a genuine one. No sooner did the instrument strike his sight than his face brightened up and he said: "Everybody can see at once that there must be tone in this," and after playing upon it he was so pleased that he said to Gemünder: "Yes, there are none of the present violin makers who have brought it so far!" Hereupon Gemünder informed him that this was also a new violin of his making. Scarcely had the visitor heard this, when, ashamed of his prejudice, he took his hat and went away.

Similar incidents often occur. In 1859 Gemünder sent violins to the Exhibition of Baltimore, after which, on one occasion, he was invited to a soiree at which his violins were played. He also had a genuine Guarnerius among his own instruments. An amateur, Mr. Gibson, a very good player, was present and anxious to hear the Italian violin. During the performance of a quartet on the violins made by Gemünder, this amateur, who was possessed of the popular prejudice against new instruments, and who fancied he heard the Italian violin, was so exceedingly delighted with it that he observed, "To hear such violins is sufficient to keep any one from ever touching new ones." But when Gemünder told him they were new ones made by him, the amateur stared at him as much as to say, "Do you make fun of me? These violins do not look new at all!" Gemünder, however, convinced him of the truth of his assertion. This fact surprised the amateur to such a degree that he was at loss what to say, and later, upon learning the price of one of the instruments, bought it. Sometime after this he valued it at two thousand dollars in gold. Since then the violin has been sent several times to Gemünder, either for a new bridge or other slight repairs, and each time new anecdotes have been related of it. Of especial interest is that one of Father Urso, who was looking for a genuine Guarnerius to give to his daughter Camilla, the celebrated violinist. He took Professor Simon with him to see the instrument. Both were very much surprised at it, not only on account of its undoubted genuineness, but also that it was kept so well. Gemünder then let them know that he had perpetrated a joke, and that the instrument was made by himself.

One day Mr. Poznanski, from Charleston, S. C., in company with his son, who was already an artist on the violin, visited Gemünder. Although still young, his father intended to send him to Vieuxtemps for his further artistic accomplishment, and with this purpose in view he was willing to buy an Italian violin. As Gemünder had none on hand, he showed him a new violin, but Poznanski declared that he would not buy a new one. Gemünder then showed him an imitation, as if it were a genuine original. The son played on it, and both father and son were highly satisfied with it; they expressed their wish to buy it and asked the price, which was given as five hundred dollars. When Poznanski was about to pay down the money, Gemünder told him that this instrument was also new. Whereupon Poznanski replied in an excited tone, "Have you not heard that we do not want a new violin?" and they left the Atelier!

When Vieuxtemps left America, in 1858, Poznanski's son went with him to finish his studies under his direction. After the lapse of eight years he returned an accomplished artist, and visited Gemünder again. He then remarked that he wished to find an Italian violin of first class, and asked Gemünder if he had something of that kind in his possession? Here he took the opportunity to remind Gemünder of the time when he had deceived both him and his father, observing at the same time very naively: "But now, Gemünder, you cannot deceive me. I obtained thorough knowledge of imitations at Paris, and also a knowledge of the genuine Italian violins, for I had an opportunity to see many of those made by the masters." Gemünder told him that he had two Joseph Guarnerius violins of first class in his possession, and laid them before him. Poznanski expressed his astonishment to find such rarities. After a thorough examination Poznanski declared there was no doubt in regard to their genuineness! He tried both violins, and soon evinced his predilection for one of them, which he wished to buy, and inquired the price. Gemünder offered each of them at one thousand dollars, but at the same time told him that he had deceived him for a second time, for the instrument which he had picked out was new and made by himself, whilst the other was genuine. Poznanski, however, told Gemünder that he could not deceive him, that it was not possible to produce an instrument like that. At this moment two friends of Gemünder, who were acquainted with his instruments, entered the shop, and Gemünder asked them in the presence of the young artist, at the same time pointing to the instrument selected by Poznanski, "who made this violin?" They replied that the maker of it was Gemünder. This appeared to him impossible, but, after deliberating on the subject, he said, "I must believe it now, and yet I don't believe it!" A few days later, becoming fully assured that the instrument to which he had taken a fancy was not an Italian violin, he bought the genuine one, which, however, was an excellent instrument, thus giving up the one to which he had first given preference. This is another striking proof of prejudice.

After a time, however, when Poznanski felt more at home at Gemünder's, he found out that the instruments made by Gemünder were the only true concert violins, and disposing of his Guarnerius, he bought a Maggini made by Gemünder; he now saw the full extent of his prejudice, and was most severe in his denunciation of all who thought that there were no other violins but the Italian to be played upon.

If Wienawski had not been seized with such a strange fancy, and had had more confidence in other artists, he would not have been compelled to change violins every now and then, for he was constantly buying one Italian violin after another and finding none to suit him, merely because none would do but an Italian instrument. Thus he came to America and played on his Stradivarius violin, which had a splendid tone in a room, but when played upon in a concert hall proved a great deal too weak, especially on the G string, when it was overstrained. He then bought one of the finest Guarnerius violins in Brooklyn, but as it did not prove any better than the other, he returned it.

To find Italian violins fit to produce a sufficient effect in large concert halls is a great rarity, since they have been mostly spoiled by "fiddle-patchers," or had not from the very beginning the proper construction for the giving out of tone sufficient to fill such halls. On just such powerless violins Vieuxtemps performed at his concerts on his last tour through America.

One day Gemünder made the acquaintance of Mario, the greatest Italian connoisseur of violins, who was decorated for this knowledge when he was at New York. Gemünder asked him to come to his shop, as he had several violins which he would like to show him, in order to have him judge if they were really genuine instruments. Mario came and viewed the violins shown to him by Gemünder minutely, nay, even took a magnifying glass to examine the varnish, whereupon he declared to Gemünder that they were genuine instruments. But the fact is they were violins made by Gemünder!

In the beginning of 1860 Gemünder was often visited by an amateur named Messing, who wished to find a good Italian violin, for he manifested an aversion toward Gemünder's productions, owing to his prejudice against new violins. At the same time Gemünder had as an apprentice a nephew, who, when he had not yet been fully three years with him, was engaged to make his first violin, according to form of Stradivarius. When it was finished Gemünder made him a present of it, and said he would varnish it so as to look old. Afterward his apprentice gave it to a friend in New York to sell it for him. This friend published in the newspapers that he had a Stradivarius to sell. Mr. Messing was the first to make inquiries about it, and bought it, highly rejoiced at having a Stradivarius at last. He then had it examined by the violin maker Mercier, in New York, who confirmed the claim of originality. Mr. Messing then went to Europe, and at Paris he wished to hear what the violin maker Gand would say, and the latter also declared it was an old instrument, adding, however, that in order to be quite sure whether it was a genuine instrument or not it would require more time than he could apply to it just then. When he went to Berlin, he showed his instrument to the violin maker Grimm, that he might hear from him his opinion as to its genuineness. Grimm refrained from uttering his opinion, yet he offered him a high price for the instrument, which the owner considered to be sufficient evidence that he possessed something extraordinary, and to warrant him in keeping his violin. After the lapse of four years, when Messing had returned to New York, he came to see Gemünder, full of joyous anxiety to show him his violin, saying, "Here, Mr. Gemünder, I have something to show you ; I have found what I have been so long looking for!" Mr. Messing then opened his box, and Gemünder, catching a glimpse of the violin, exclaimed, "That is my apprentice's first production; how did you come by it ?" At these words Mr. Messing stood as if thunderstruck, and in his bewilderment he tried in every way to convince Gemünder that he was mistaken, but failing in this attempt, his discomfiture was complete. When he had somewhat recovered from his dismay, he felt heartily ashamed, because he had disregarded the work of the master only to take up with the apprentice's first production, and this, too, under the delusion that that work was a genuine Stradivarius violin. Mr. Messing is now cured of his prejudice, and is no longer looking for a Stradivarius violin.

At the time when Gemünder had his violin in the Exhibition of Vienna, Baron Leonard, from Hungary, who was a great violinist, brought him his Italian violin to have it repaired. During their discourse about violins the Baron conveyed to Gemünder the impression that he had already seen many Italian violins, and he seemed to have a great knowledge of them. Thereupon Gemünder showed him a violin that seemed to be a genuine Guarnerius, which he had determined to send to the exhibition of Vienna. The Baron was quite astonished at seeing such a wonderful and splendid instrument, and did not know which to admire more, whether the varnish of the violin or its tone ; in short, he looked at it with reverence, as if it were a shrine. Gemünder then showed him a Stradivarius, and when the Baron's gaze fell upon this instrument, he seemed to be enraptured, and he exclaimed, in a tone of question : " Mr. Gemünder, how do you come by such treasures ?" In truth you have a treasure of the greatest rarity, for I never saw a violin so beautiful and of such tone !" When, however, Gemünder declared to him that these were the sisters of the " Kaiser " violin, which was in the Vienna Exhibition, and were made by him, the Baron conducted himself as if he had awakened from a sweet dream, and found it difficult to realize his true condition.

PREFACE.

It is not my intention to unfold in this work my knowledge of the structure of violins; for the present generation would not thank me for doing so. In the treatise itself will be found the reasons why I have not set forth that knowledge. Since the death of the celebrated old Italian violin makers, many works have been put forth, in which we find not only in what manner those famous masters varnished their violins, but also prescriptions even, of theorists who usually know nothing about the practice, or mathematical principles thereof. Abundant theories are to be found in all such works, but they are good only for those who have little or no knowledge of violin making. If the science of the celebrated Italian masters could really have been found in these works, the experiments made by European investigators would not have been entirely unsuccessful.

In George Hart's interesting book, " The

Violin," a comparative illustration may be found of the workmanship of all violin makers with whom he became acquainted, either personally or by history, and by whose productions he obtained his practical knowledge, which comparisons are generally good, but not entirely free from error. This compilation of experiences is highly interesting for all those who take an interest in violins. The treatises which will be found below have reference simply to the art of making violins, to violin players and their critics, the information contained in which has for the most part never hitherto been made public.

Through these scientific explanations a better judgment will be awakened, which will tend to show how, in consequence of mistakes and ignorance in regard to violins and violin makers, false ideas arise.

PROGRESS OF THE STRUCTURE OF VIOLINS—THEIR CRITICS.

In 1845 I became personally acquainted with Ole Bull, at Vuillaume's, in Paris, where I then had my first opportunity of hearing and admiring an artist on the violin. I learned then to appreciate the beauty of both arts, and the sublimity of attainment in either to be a violin virtuoso or a perfect volin maker. The latter art engaged my whole attention, and it was my greatest aim to reach to the highest point of perfection therein.

I also found that Ole Bull took special interest in the different forms of violins, and I remember that as early as 1841, at which time I worked at Pesth, my employer made the so-called "Ole Bull's bass-bars" in violins, the ideas of Ole Bull concerning violins then being accepted as authority. Ole Bull subsequently made many experiments regarding tone, especially upon new violins, in order to reproduce the same character of tone, then considered lost, peculiar to the Italian instruments. Knowing that all experiments made since the death of the celebrated Italian masters had proven unsuccessful, he undertook to construct a violin of very old wood, but was soon convinced that he had not obtained better results than others; he therefore decided the project to be an impossibility, and having arrived at this decision, his opinion was generally conceded to. Since then, doubtless, he found out that to make a violin was a more difficult task, for him, than to play on one. As a virtuoso, however, he obtained a celebrity which will make his name immortal, and as he was an artist in his own peculiar way, his name will live forever in the memory of men. Nature has endowed many men with rare gifts, each one possessing a talent peculiar to himself; but we know how long it requires to perfect one's self in any given art, and it therefore cannot be expected that a great violin virtuoso should at the same time be proficient in the art of violin making, the two arts being totally different. It is, however, generally believed that the assertion of Ole Bull had more weight with many violin players and amateurs than the most adequate knowledge of a violin maker. I admit that Ole Bull had some experience with violins, but had he obtained sufficient knowledge he would have easily understood that many of his ideas were not based upon principles which he thought had remained secret to all investigators on the subject, as the greatest authorities have acknowledged the tone in George Gemünder's violins to be of the same quality as that characteristic of the best Italian instruments.

This proves that violins are judged the best when they are mistaken for Italian instruments and prejudice only is the actuating motive when the declaration follows that the instrument is a new violin. If, therefore, the knowledge of tone could have proved more reliable, prejudice would not, in many cases, have appeared so severe, and embodied itself so as to degenerate into fanaticism.

Violins made of healthy wood and according to the rule can never lose their tone. It is, however, something different if they are carelessly treated.

When an Italian violin, which lay untouched in concealment for fifty years, was shown to Wieniawski at the Russian court, and he was asked what he thought of it, he said, after trying it: "The violin has a bad tone." "Well," said the Emperor, "let us put it back in its old place. If it had been good I should have presented you with it." Wieniawski, greatly surprised, replied: "Oh, when I play upon it it will regain its tone." Here vanity and ignorance are shown at once; for if that artist had had any knowledge of violins, he must have known that the violin was not in good order, and that it was first necessary to have it put in a good condition by a professional repairer; but instead of making such a proposal, he thought to make an impression by his renown, and that he would improve it by playing upon it.

I mention this because it contains two points: firstly, because, especially here in America, great stress is laid upon the opinions of such artists, but it proves that artists do not always have a knowledge sufficient to enable them to give a correct judgment of violins; secondly, if this violin had been new, many would have thought that it was made of chemically prepared wood. A violin, however, of such defective wood, can never give a good tone ; because the life is taken out of it when it is made. If such artists would make themselves acquainted with a professional violin maker, many of them would get more light on this matter, but since they consider themselves to be authorities on the subject, there is very little prospect of visible progress. It is, therefore, a rarity when an artist is found who is able to judge of the quality of tone, whether the wood is chemically prepared or not, and although this is easily to be distinguished by the practiced ear, a peculiar experience is required for it nevertheless. Many, however, believe that he who plays the violin to perfection, and especially the player of renown, must be acknowledged as a judge of tone. I admit that many violin players are judges of tone, but not beyond a certain degree, as the greater number of them hear their own instruments only and are taken with them ; but he who possesses a feeling of tone, and into whose hands violins of all shapes and qualities are falling, whereby he learns to distinguish the different characters of tone, is to be considered a connoisseur of tone ; he must, however, possess some knowledge of playing, although it is not necessary for him to be a solo player, for with how many solo players have I become acquainted who have no more judgment of tone than children.

For musicians and solo players it is very difficult to find out how far the tone of a violin reaches. Many a player, having no experience in this regard, plays in concerts on a violin which sounds like an echo, but if the instrument is called Stradivarius or Guarnerius and $3,000 has been paid for it, and besides it has a "history" attached to it, then, verily, it must sound. The critic, however, does not blame the violin, but the player, for weakness of tone, and in that respect he is right.

For solo players who still use such echoing violins in concerts, it would be of the greatest importance to make themselves acquainted with the quality of tone which is fit for concerts, for most Italian violins which are used in concerts prove either too old or of too thin wood; but most players are accustomed to the fine, tender, echoing tone to a degree that the true concert tone appears quite strange to them.

Thus, violins of chemically prepared wood

will never do for concerts, and it is a great mistake to believe that such violins have ever produced as good a tone as good Italian violins do. Ignorance and self interest have launched this untruth into the world. For violins made of such wood produce short vibrations—a muffled color of tone similar to that of impaired Italian instruments. Vuillaume put all the world in commotion with his violins of chemically prepared wood, and all the world sang hosannas. But when it was found that such instruments kept this tone only a short time, there arose a general prejudice against new violins and no one would play on them.

In order to remove all such ideas and prejudices I can safely assert that violins of a free, high, clear and powerful character of tone, with a quality which thrills the heart—such tone as my instruments produce, and which qualities are now seldom found in the best Italian violins—can never be obtained by any artificial preparation of the wood, but only by way of science according to acoustic principles.

Of course it is the wood more than anything else which is to be taken into consideration;

for without the right sort of wood all science will be unavailing, and *vice versa*. Many violin makers can get the best wood, but where there is no talent applied in the construction, nothing very good can come forth.

Of all productions of art, the violin is the most difficult to judge, and I have nearer illustrated the different characters of tone which violins produce, and tried to make these things more comprehensible, in order that this medley of opinions and judgments which have been given may be put in a clearer light.

I was highly astonished at the manner in which my "Emperor" violin ("Kaiser" violin) was judged, which was sent to the Exhibition of Vienna three weeks after it had been finished. The violin had attracted not only many admirers, but also a great number of gazers who have no idea of a violin, and who stared at it only on account of its price.

Thus, the New York *Staats Zeitung* had a correspondent in Vienna, who also stared at the violin from the same reason. His ignorance, which he exposed in his correspondence to the newspaper which he represented, led him to make the following remark, which was published on the 27th of June, 1873, and runs as follows: "From Salzburg several violins, mostly the former property of Mozart and Beethoven, were sent, and the one which Beethoven owned was made by Hellmer, at Prague, in 1737, as was noted on the label, (saleable for 200 Florins,) while for a Gemünder violin in the American division of the Industrial Palace, $10,000 (!) are asked. Of course, everybody laughs at the simpleton who believes this is the only curiosity of the kind, and thinks he can obtain such a fabulous price for it. The Commission that for this time has made us very ridiculous with our "Go ahead," should remove that label as soon as possible, that one of the exhibitors may not become a public laughing stock." But that writer soon found how much this violin was admired; he learned to see that it was the only curiosity of the kind, in faƈt, for soon afterward I read again in the *Sontag's Staats Zeitung* that "the violin was admired very much."

This violin was exhibited by me for the purpose of proving to the world that I can make violins that have the tone which has been sought for a long time since the death of the celebrated Italian masters, since which all attempts have miscarried, and I confirmed this fact in a circular added to it.

But what was the result?. It was not believed. In the Exhibition of Vienna my violin was mistaken for a genuine Cremonese violin, not only for its tone, but for its outer appearance, which was so striking an imitation according to Joseph Guarnerius, that a newspaper of Vienna made the observation: "George Gemünder cannot make us Germans believe that the violin sent by him is new; a bold Yankee only can put his name in a genuine instrument, in order to make himself renowned!"

Although this was the highest prize which a violin maker had ever obtained, it was no advantage either for me or the public; for the art of violin making was not furthered by it, but rather still more impaired by the correspondence of the *Staats Zeitung* and the New York *Bellestristic Journal*. The latter writes as follows: " S. F., Pittsburg.—G. is a pupil of Vuilliaume; his violins are much demanded, but their prices are so high that purchasers are frightened!"

Thirty years ago I sold violins at from $50 to $75; ten years ago I sold violins at from

$100 to $300; now I sell them at $100 and upwards; and violin makers here and in Europe ask the same prices. Nay, amateurs who do best in their ignorance, ask still higher prices. Wherein, therefore, do we find that which frightens the purchasers? The effrontery of writers who make such statements as the above will bring them no honor.

Many may still remember that I had determined to send six violins of different forms, copies of the best old master-violins, to the Vienna Exhibition, and intended myself to take the matter in hand, but, owing to an accident, I was compelled to give up this intention. In consequence, I resolved to send only one violin. To select one of them, artists such as Wollenhaupt, Dr. Damrosch, Carl Feinninger and others were consulted, but they differed in their opinions, which may be taken as a proof that the instruments were very much alike in character; they are also witnesses of the fact that I made them. In order to call attention to the one selected, I noted the price " ten thousand dollars!" Nobody, however, was charged to dispose of it, although three thousand dollars were offered.

The circumstances connected with the construction of this violin gives it more than an ordinary interest. Ridicule and praise in the highest degree are interwoven with its history; therefore, it has been hitherto the most interesting new violin in this century. Why I could not be its representative and had to leave it to fate can be learned from what I have already written about it, and how I have judged every thing connected with it. I was, however, sure of one fact, namely, that it would be acknowledged as a production of art. The admission must then be made, and the claim is amply justified by facts, that, as new violins are frequently mistaken for genuine Italian instruments, even when most particular attention is given to the varnish, the art of violin making must no longer be considered as a lost one.

May the foregoing satisfy all doubters and those who have lately, especially in America, written about the lost art of varnish and tone, and may it cause them in future to refrain from investigating into the so called lost arts. He who would give a scientific explanation of this art and be a critic, must be thoroughly acquainted with it.

A TREATISE UPON THE MANNER IN WHICH MASTER-VIOLINS ARE RUINED.

The manner in which violins are so often ruined seems almost beyond comprehension, or rather the way they are generally treated must necessarily involve their ruin. The cause of this can not be entirely ascribed to those destroyers of violins who pretend to be repairers, but it generally rests with the owners of violins themselves, because they are usually ignorant as to who is master of the art of violin making and to whom a master violin may be entrusted. They therefore make inquiries for such experts, and apply for that purpose, generally, to renowned violin players, not realizing that even these are not always endowed with discrimination, frequently not more so than the one asking advice, and thus the latter is led astray.

To find an adept repairer is as difficult as to find a thorough master of the art of making violins; for the repairer must possess the same knowledge of the production of tone as the best violin maker. The man who cannot make excellent violins cannot be an excellent repairer. To obviate all doubts on the subject, I will state that the foundation of the whole secret is simply this "Every violin maker will make repairs in accordance with his knowledge, as he would make violins, and violins as he would make repairs!" This principle is so scientifically correct as to be conceded even by the most severe critics.

Many a man achieves a reputation by certain meritorious accomplishments in which he has distinguished himself, and in consequence thereof everyone believes him an artist in the fullest meaning of the word. For instance, Ludwig Bausch, of Leipsig, gained a deserved and world wide celebrity as an artist in making bows. I also esteemed him as an excellent and very accurate worker. But to my astonishment I found, as I regret to say, that his fine repairs were mostly devoid of value, as also were his new violins, so far as the production of tone was concerned. But artists and amateurs, far and near, adored his useless repairs and new violins, which latter usually sold for high prices.

Thus the public are unable to form a proper

judgment in regard to the art. It would pain many a one, if they could realize the manner in which valuable violins are treated by such violin makers and repairers. Repairing violins, therefore, is as little understood as violins themselves, in consequence of which not only the interior of many an Italian instrument is ruined, but also the exterior is often deprived of its classical appearance by an alcoholic varnish, which is smeared over it and which impairs its value ; and yet many owners of such instruments, who do not know any better, rejoice to see their violins with such a glossy surface.

To rehabilitate a valuable instrument, and repair the exterior if necessary, requires a skill as artistic as the rehabilitation of a painting by a celebrated painter. Such instruments are also often peculiarly tortured by unskilled hands, and many a valuable top has been damaged by the operation of putting, or rather forcing, in the sounding post.

Owners of violins should take particular precaution never to permit the cutting away of wood out of the bottom or top of a violin, without being fully satisfied that the repairer is an adept in the art. In Italian violins made by the old celebrated masters there is no necessity at all for doing this, as they have not as a rule any too much wood, and most of them are poor enough in this respect; in case those artists made no mistakes others have brought them in by their repairs.

OF THE CAUSES WHY THE JUDGING OF VIOLINS AND THE REPAIRING OF THEM IS LESS UNDERSTOOD THAN OTHER ART PRODUCTIONS.

Beautiful and interesting as is this art of making and repairing violins, and however great has been my enthusiastic devotion to it, I should never have engaged in it had I in starting possessed my present experience, for the ignorance which the public has shown by the confusion of opinions in this branch might almost make one believe these judgments emanated from a mad-house.

Why is it we hear no such conflicting opinions about the productions of any other branch of industry or art? Because in no other business do we find so many pretenders. And why is it they infest this particular branch of business more than any other? Simply because the art of violin making is not founded on a correct system, and this may account for the medley of ideas which have been spread broadcast throughout the musical countries, except France, where a regular system is recognized.

Yet in spite of the lack of correct system of making violins, I have become acquainted with a few German musicians who have acquired an excellent schooling in the art. In this respect I cannot refrain from mentioning my admiration for a thoroughly skilled musician, Mr. Herman Eckhardt, of Columbus, Ohio, a man of rare genius in the knowledge of music, who was able to define clearly and accurately the different periods of the progress I made in violin making.

Such a man I must respect the more, because he is endowed with sound judgment, which other musicians, often of very high standing, could only acquire by instruction, a method which to some of them would seem to be impossible, as they are devoid of judgment, having their ability warped by false ideas about violins, and rendering them incapable of correctly understanding and appreciating the latest and best productions; this may account for their fanatical admiration of Italian violins, even if they possess only imitation, but, as "ignorance is bliss," they are happy.

On the other hand, there are amateurs who

take such a practical view of the matter that they are just opposite in their beliefs to this class of fanatics. They do not see why a new production, which answers the purpose as well and which in more ways than one is preferable to an old production of the same kind, should be regarded as of less value. They do not understand why a desirable article should command an enormous price when another article accomplishing the same effect can be bought much cheaper. And in this they show a common sense which might well be emulated by many others. While it is true that an enthusiast ought never to be blamed for his enthusiasm, if it has a reasonable base, it is no less true that lacking in this respect he is nothing more or less than a fanatic. This class of people is by no means exclusively confined to amateurs, but even includes in its ranks many true artists in music.

ON THE PRESERVATION OF VIOLINS.

There is no doubt that a certain class of violin players pay very little attention to the care of their instruments, as they use them daily, and few have time to bestow the necessary attention upon them. If a violin is out of order, a musician or amateur who knows nothing about it continues to play upon it. At length he perceives that the tone is not the same as it was before. Many, therefore, often lay the blame on the repairer, or on the violin maker, if it is a new instrument. It is therefore desirable that players should always pay attention to their instruments and examine them whenever they intend to use them, to see whether everything is in order; that the neck has not sunk a little to the front, causing the finger board to lie deeper on the top and the strings to lie somewhat too high. Such deviations will occur, particularly when the top is very much vaulted, as well as by change of weather or climate.

As soon as the weather becomes moist it is

advisable to keep a violin in a box ; when the weather is fine it should be taken out of the box for a time every day ; and even if it is a very old violin it is not good to keep it always locked up. A violin should never lie on a floor, whether in a box or not, but should always be kept on an elevated place and in a moderately warm temperature.

Before using the violin it is advisable to rub it with a soft cloth or chamois, so that neither dust nor perspiration may remain on it ; it should also be cleaned each time after being played upon. The sounding post should also be examined, to be sure that it still stands perpendicular. The bridge, too, must be looked at, and if it stands obliquely it must be brought into its normal position again before taking the bow. It usually inclines somewhat forward on the E string after tuning it. If this is the case, pinch the E string between the thumb and index finger, while the corresponding part of the bridge is moved backward by the points of the fingers.

On good and excellent violins particular attention must be given to the bridge, especially when it fits the instrument, for it is not always easy to replace it with one equally good. A bridge which is qualified to affect the violin and contribute to the charm of tone of the instrument is more valuable than one would often think. Many consider a bridge of as little consequence as a string, when it breaks on the violin, and think they can restore the loss by a bridge which costs three cents; for the correct model of a bridge is considered only as an ornament by such people. Of course they do not know that this is one of the most important parts of good violins, and that there are but few violin makers who are able to make a bridge as it should be. But it is the same with the bridge as with the violin.

It is not only the correct construction of the violin and bridge which produces a good tone, but the right sort of wood must be found for the purpose. Thus the bad form of a bridge made of fine wood is just the same as a common fiddle made of fine materials. It therefore follows that we should take as much care of a master bridge as of the violin itself.

It some times occurs that the sounding post of the violin becomes shorter by itself; in this case it may be advisable to relax the strings entirely in order to see whether the sounding post does not fall. If this is the case, a new one must be made of old wood by a skilled workman. The cause of this is that the wood contracts more or less, especially in dry weather ; this may also be caused by a change of air, which sometimes even produces a distortion of the swell of the top.

When such care is habitually taken, a violin will always be in good order. Too low a sounding post causes a lower position of the top on that side, which, when not remedied, will remain and will produce a defect in the swell and tone. This is also the case when the sounding post is too high, and many violins are seen where the swell is higher or lower than it ought to be on the side where the sounding post stands. This is also the case with the bass-bar or so-called " soul" of a violin, which is just as mysterious a part of the violin as any one can imagine ; and its quality shows the skill or ignorance of its maker.

TO ILLUSTRATE HOW VIOLINS OF MY CONSTRUCTION MAY BE COMPARED WITH THE OLD ITALIAN MASTER-VIOLINS.

From the foregoing treatises it will be seen with what energy I devoted myself to the art of making violins, and I can declare to the world with a good conscience that I have reached the standpoint in this art which has been striven for in vain during a century.

I have studied all the characteristics in the construction of the Italian master violins, and have had extensive practice in imitating violins, as masters have made them, and have obtained an understanding which enables me to unite all good qualities of tone in the construction.

As I am able to judge from experience, nobody can confute me. All those who doubt it or will dispute it can neither confute me scientifically, nor prove what they say. I have had a great many opportunities to hear and repair the best Italian violins myself, including Paganini's wondrous violin at Vuilliaume's, in

PROGRESS IN VIOLIN MAKING.

Paris, and I can affirm that my "Kaiser" violin can be considered as wondrous a violin in regard to tone and character as—nay, it is even to be preferred to—that of Paganini's in many respects!

I also make a peculiar kind of Maggini violin. For this purpose I have selected an older form than that which is generally known. I construct these violins in a manner to include all good qualities of tone, and they are, therefore, far preferable, because they surpass those of Stradivarius in greatness of tone. Such distinctions prove that I have made great progress in this art.

Most Italian violins are now of interest only to admirers of art, and may be recommended to antiquarians, for there are only a very few still existing which can be used for concerts, and although if even their voice disappears more and more out of their body, they will always be valued, kept as relics and admired by friends of art. But it is only fancy which makes most of them adore what they do not understand, and they trample down the blossom of the new productions which the world brings forth.

Therefore, it will be of some interest to

many to hear more minute particulars about the method of construction of violins of the old Italian masters, as many persons are still in darkness as to which violins the best tone is to be ascribed. This want of knowledge comes simply from the fact that a combination of uninjured instruments of the best masters is a task very difficult to be effected, and these instruments would by all means have to be put in proper condition by an expert.

This has, perhaps, never been done yet, and a general comparison could only be made as the opportunity presented itself.

As I acquired knowledge of the system, the forms and swells of violins of the great masters, I also became so thoroughly familiar with the characteristics of tone that I have found out what the present needs require.

I will now consider in detail the different characteristics of tone of the productions of the great masters, and state in what manner this difference was obtained.

Jacob Stainer, at Absam, in Tyrol, was a pupil of Nicholas Amati, at Cremona. Stainer and Amati made violins which were mostly demanded by amateurs on account of their round, sweet, silver tone. This character of tone they produced by a small, round and somewhat oblong swell, as well as by a neat and somewhat smaller size than that of Stradivarius, who endeavored to gain a greater sonority of tone. Stradivarius, therefore, made the swell less high than Stainer or Amati, but of a broader circumference, drawn oblong, by which he obtained a sublime tone in an aristocratic and majestic form.

Joseph Guarnerius del Jesu.—As long as he made violins according to the school of his great master, Stradivarius, his productions were of a similar nature. Later, he made somewhat smaller models, sometimes with a circumferential swell, by which he gained a somewhat smaller tone, but with a striking, quick touch of a peculiar brilliancy. It is strange that he gave a different form to each of his violins, the f, the swells and the scrolls varying in almost every instrument. It is told that he was imprisoned for a long time, and, under great deprivations, he made violins secretly. In all his productions his great genius is recognized.

Duffu Prugar, at Bonninien, lived in the sixteenth century. His violins have a large and wide form, with interesting ornaments of carving work and inlay; their swells are beautiful, and as high as those of Stradivarius, and they produce a great and full tone. But as there are only few still existing, many violins are imitated in France according to this model, and they are spread far and wide.

Maggini's violins are mostly of a large size and of a higher swell and fuller toward the extreme parts than all the other violins of the Italian masters, therein producing a great fulness of tone; on the G and D strings their color of tone is particularly deep.

Gaspard da Salo made very interesting violins of small and large size; the former have a peculiar character of tone, not very strong but of a very clear color. These violins have a beautiful, high and round swell, similar to those of Jacobus Stainer, but those of a greater size are flatter, producing more power of tone, and are therefore better adapted for solo performances.

These celebrated masters left us a great choice of different forms and swells, as well as their method of workmanship in regard to the top and bottom of their violins, where the proof is to be seen that they always made investigations in order to gain a greater perfection. Stradivarius and Joseph Guarnerius have especially obtained a beautiful quality of tone in their violins, yet in order to gain an easy touch of tone, they worked the top pretty tender, and in many instances they made the middle part of the top most thin, probably to further the easiness of sound still more. Such violins do not answer for concerts.

It seems that at that time less attention was paid to such a power of tone as is required now, because only few of them have been found with an acceptable thickness of wood in the top and bottom. This is, therefore, the reason that so many Italian violins produce too weak a tone in concerts.

Although Maggini left the top and bottom thicker in the middle part, still, most of his violins have not, on account of construction and deep color of tone, been received with favor like those of Stradivarius and Joseph Guarnerius. As only a few such Guarnerius and Stradivarius violins were found which by reason of their thickness of wood answered the purpose of solo violins, every one believed all their productions of a like character.

Therefore, so many solo players often expose

their ignorance by playing on such violins in concerts.

Stradivarius instructed other pupils besides Joseph Guarnerius, who made excellent violins, and many of these violins still exist. As the most of them were made with the full thickness of wood, they produce a splendid tone, often better than some of those made by their great master. This teaches us that he who wishes to possess an Italian violin on account of its tone cannot depend upon finding it by the name alone, but he has to pay all his attention to the discovery of those in which the necessary thickness of wood is found.

A solo player, therefore, should never play a violin on account of its name alone, for if the violin produces a weak tone, the blame will be laid on him, and so much the more because it is generally supposed that such instruments must be master violins.

ART EXHIBITIONS.—HOW VIOLINS ARE EXAMINED AND JUDGED.

First of all I will take America into consideration, where the art of making violins is too little understood to be judged. Commissioners of exhibitions like those, for instance, of the late Centennial, have no idea of violins, and, therefore, are unable to appoint judges competent to award the premiums. It would be too much to ask that they should themselves be such connoisseurs, for the violin is still considered as a fiddle in this country, and it may still take a long time before the people here reach the standard of knowledge and appreciation which Europe occupies. Therefore, only very few real violin makers are found here, for most of them are only amateurs doing business in this branch. In the Centennial exhibition in Philadelphia, in the United States Department, were found mostly such amateur violins. I have heard that all those who called themselves violin makers received a premium. The judges were either unequal to the requirements of their office or they desired to offend nobody. If the latter be the case they certainly acted generously if not justly. But exhibitions of art were established for the purpose of finding out in which way the different articles of industry and art compare with each other. Proper examinations can be made only by professional men, otherwise only that fiddle that "cries" the most will attract the greatest attention.

Justice will never prevail in such exhibitions, owing either to want of knowledge in order to be able to judge who has deserved a premium, or to favoritism, for merit can hope least, especially in Europe. Artists there can only receive acknowledgment if they have the means to spend. The Centennial exhibition, however, was not guilty of such a wrong; here it was the desire to be as just as possible to all, although not every one could be satisfied. To act in the capacity of an awarder is always a thankless task; whether the judge has or has not the necessary knowledge, discontent is sure to follow, because the conceited man who has been unrewarded does not see the difference between his production and the better one of his co-exhibitor, but an injustice is done to an artist, if through favoritism a premium is awarded to an inferior production.

Exhibitions, however estimable they may be, are still very imperfect in regard to their organization; in Europe they have been for years entirely corrupt, and are now called into existence mostly by speculators. The true principle has been lost sight of and taken a corrupt form. It is scarcely to be expected that the time will come when the many defects which have crept in will be removed again, for all these failings which have manifested themselves throw a shade over such exhibitions, and the time is not far distant when they will be entirely disregarded, if not reorganized on a different basis. But I believe that they will never attain great perfection, even if taken in hand by the Government, for so long as a system of awards is connected therewith, mistakes and discontent cannot be avoided. Managers of exhibitions are not always competent to appoint the proper professional men and experts as judges; and as those appointed lack the necessary qualifications, dissatisfaction ensues. But suppose the awards were made with proper knowledge and strictest impartiality, what then? What have the remaining competitors gained who are less gifted by nature, and therefore could not receive any award? Nothing but mortification and an impaired business. Is this fair on the part of human society? Not every one can be an artist. The offering of premiums has for its object the promotion of industry; but the majority of exhibitors can never achieve distinction by reason of lack of talent, and must consequently be considered as excluded from their line of business. Are we not bound to consider them as our fellow brethren and to care for them as well as for those receiving premiums? But the present generation does not seem to have any thoughts about this, for there are but very few men who are still animated with noble impulses; while the majority are striving to ruin their fellow men by greediness.

In my opinion such exhibitions cannot continue any longer, because justice can never be expected, and the chase for the highest premium in order to outdo others, has not only become ridiculous, but also immoral.

If I were the richest man, it should never come into my mind to strive for a premium which I must purchase through so-called leeches. There are, however, connoisseurs who know how to distinguish that which is better from that which is less good.

As long as such exhibitions are based on such rotten principles, I find no longer any interest as an exhibitor in striving for a premium, and as I gained the highest moral premium in the exhibition at Vienna in 1873, on this account I did not compete for any premium as an exhibitor in the Centennial exhibition at Philadelphia!

NOTE ABOUT DILETTANTI VIOLIN MAKERS.

Whoever takes an interest in violin making will undoubtedly be pleased to hear more particulars in regard to dilettanti violin makers and their patrons. There are some dilettanti violin makers in America who consider violin making their business, and there are others who do not make it their chief business. They have their own particular patrons, who in the knowledge of violins are on the same level with themselves ; but it cannot be denied that in the productions of some of these violin makers there is talent discernable ; if these persons could have had proper instruction, more good violin makers would be found than are now in existence. But as long as dilettanti violin makers remain as such, only dilettanti violins will be produced ; for without proper instruction it is impossible to obtain either a correct knowledge of the exterior formation or a correct knowledge of the production of tone.

It is true, that every piece of wood over

which strings have been stretched will sound, and every such instrument will have its admirers. There are, however, dilettanti violin makers whose self-conceit and boldness is simply astonishing. The professional will understand this, for if a self-conceited man could see clearly and look into the matter, he would be astonished at his workmanship, as I was once myself.

As dilettanti usually lack that practice which is peculiar to the regular violin makers, they very often experiment in all kinds of machines by which they expect to lighten manual labor; their object, however, is mostly reached in a very roundabout manner, although they believe to have made an improvement, and this improvement they announce to the public as a great success. As most of their patrons have no knowledge of the matter, such a dilettante appears to them as an extraordinary genius. This supposition would perhaps not be disputed if it did not take considerably more time to execute with their machines a certain amount of work than the practical workman requires simply by the dexterity of his hand.

A dilettante violin maker can never be a thorough workman, and is entitled to be considered only as a "jack-of-all-trades;" he has a great many kinds of tools which the regular violin maker never uses.

Many dilettanti are presumptuous enough to believe themselves further advanced in theoretical knowledge concerning tone than the most experienced violin maker of the present day. Some of them ask, in consequence, a great deal higher price for a violin of their own make than does any regular violin maker for his. But it seems to me that such persons are often only the tools of Ole Bull, a once celebrated violinist with extravagant ideas, who misled them. They, however, believe to have learned from him the true secret of the art of violin making. He also tried to persuade them into the belief that in order to have violins sound well and to be serviceable for concerts, they ~~should be~~ *are* made of chemically prepared wood. If such pretended wise man would have some knowledge of wood, he ought to be able to distinguish wood which is chemically prepared and that which is not! About this point I have already sufficiently explained my opinion.

To give the wood the old natural color which is peculiar to the Italian violins, in a great measure depends on the material used, for not every wood intended for violin making has the necessary qualifications. Violins made from such selected wood are therefore especially valuable.

It cannot now appear strange that the general public has so little knowledge in the judging of violins, when a world renowned violinist like Ole Bull shows such ignorance. Here in America the latter preferred the company of dilettanti violin makers, for the reason that they were generally willing to listen to his ideas, and some of them have studied now so much that they cannot see any clearer nor hear any better.

Dilettanti violin makers form a peculiar class of violin makers in America; and they seem to be born for the sphere of such knowledge as is here shining forth Their patrons write articles for them in which they try to instruct the public by their ignorance, as we find, for instance, in the Philadelphia *Times*, of August 30th, 1879: "Gemünder refuses to state the source of supply for his wood, and it is a well-known fact that he and others use at times chemical preparations for the purpose of changing the character and the appearance of their wood."

The writer of this notice made a statement without any foundation. Had he and his train a proper knowledge of the matter, they would be able to perceive that the material of my violins is not chemically prepared and the character of the wood has not undergone any change whatever. It is presumptuous in ignorant persons to make such statements against a man of long experience, for the purpose of bringing his productions into discredit; productions which are proofs in themselves that not a single violin can come into the condition of those manufactured of chemically prepared wood, as those of Vuillaume in Paris. But such individuals manifest not only a prejudice against a better understanding, but also are impertinent, from which stupidity and meanness emanate; and thus they unmask themselves as false experts.

The cause for this assertion will have to be found, and for the disbeliever there is no other ground in the advantages I have gained by my studies, which to them seem impossible; and as the Italian violins are generally acknowledged the only good instruments, they try almost anything to oppose what has proven itself so gloriously, rather than acknowledge it as a fact.

Truth, however, can never be overruled, and the time will come which will impose silence on such individuals! Since mankind inhabits the earth their characters are as different as we find different plants. Many a flower is not fragrant, and how many stately and celebrated men are heartless! Those, therefore, who are void of generosity are able to do evil. Those classes who are as it were idle weeds, for the kinds are both useful and hurtful to men; all that nature produces has a meaning. If we could fathom all the secrets of nature we would also be able to understand the meaning of them, and idle weeds could be less hurtful. But in nature there lies a wisdom which remains a secret to mortal man.

GOOD LUCK AND ART, AND REMARKS ABOUT VIOLINS.

It is an incontestable fact that the success of the endeavors of men to gain a livelihood depends upon luck, although many are of different opinion, especially those who are always favored by good luck, as they ascribe their success to their enterprise and skill. They do not consider that good luck only has offered them a chance. Many become wealthy without being gifted with peculiar knowledge, while many others, in spite of all their knowledge and genius, endeavour in vain and do not see their efforts rewarded. It is, therefore, a matter of fact, that neither art nor science produce wealth, unless they are favored by good luck, and the cases are innumerable which prove this. From the many experiences in my life, especially in my profession, I will only mention the following: Vuilliaume, of Paris, was favored by nature in a very high degree in every thing; he was not only the greatest artist in his profession in Europe during the present century, but also an excellent business man, and good luck smiled on him in all his enterprises. Lupot, his partner, laid the foundation of Vuilliaume's independence by effecting a marriage between him and a very rich lady of nobility. Thus he became not only a celebrated man, but also the richest violin maker of our time. Although some of his violins of prepared wood incurred discredit, nevertheless there were admirers who bought his violins, even in America, where the prejudice against new violins is so prevalent, on account of the supposition that the wood of them was chemically prepared, a practice of which they so stupidly and unjustly accused me, and thereby caused a great deal of harm to my business. On the other hand, Vuilliaume, who really prepared his wood in a chemical manner, was lucky and prosperous.

What is the reason of this and where is it to be found, and why does good luck generally lie in the opposite extreme? The solution of this secret will probably remain undisclosed to mortals. Upon whomsoever fortune smiles, and whom she allows to blow the golden horn, he penetrates the world, his name becomes great, and he produces upon mankind that effect which persuades them into the belief that.

the best can be found only in him. If Vuilliaume had been a poor man he would have certainly remained poor, especially in America, where the art of violin making is still less understood than in Europe, and unjust reports will be more readily listened to than anywhere else.

In Europe there was a general supposition that a pretty good demand for old Italian violins existed in America, in consequence of which dealers in old and new violins found their way hither. In disposing of these instruments they were not very scrupulous in regard to the information, and sometimes gave them names according their own fancy. A great many so-called Italian violins and violoncellos came in this way to America, and the owners are happy in the imaginary possession of an Italian instrument. Other persons again entertain the idea that they are surer of a genuine article if it comes from Europe, as there is their home; but if it is believed that this is always the surer way, it is a mistake. It requires an extraordinary study to recognize the maker of an instrument, and understand the dead language of the violin. Thus it must not be believed that the instruments claimed to be

Italian are always genuine ; the seller himself may sometimes be mistaken. Many owners of such "baptized" violins do not always like to be informed of the real origin of the instrument by a person of thorough knowledge.

Sometimes I feel constrained to give an opinion by virtue of my knowledge, but it it must not be expected of me to admire a thing that is not genuine, as those owners do in their ignorance.

If, howevever, a genuine and valuable Italian violin has lost any part, and if a violin maker possesses the art to restore the missing part, either in imitating the varnish or in adapting the lost part to the character of the violin, so that the instrument reappears in its originality so completely that the connoisseur is deceived, the value of the violin is in that case not impaired. This also occurs in regard to very valuable old pictures, and the artist who is found to be able to execute such work is well paid.

Such artists are, perhaps, more to be esteemed than the maker of the original, as they are rare, especially those who are able to restore the originality of valuable old violins. The instruments lose their value in case the repairs cannot be carried out properly, owing to a want of genius upon the part of the repairer.

I have often shown this art in exceptional repairs; but what can be gained by it? The greater number of those who own violins do not know how to appreciate such skilful work, and, in their ignorance, they attempt to do harm in the bargain, when they hear that they must for such repairs, perhaps, pay somewhat more than usual—an additional proof of how great the darkness still is in judging this art. The time when a better understanding in this regard will come to daylight is still far off! And why? Because all other arts and branches of industry are based upon solid ground, as the State governments protect them, and, therefore, they can come to a proper degree of perfection. The art of making violins does not enjoy this privilege (except in France) and it hovers mostly in the fog since the death of the celebrated Italian masters.

Therefore, it can yet be called only a fancy art. The opportunity which has been given to mankind in this century to make this science general has not been regarded, because the confidence and belief in it has been wanting, and it will disappear like a drowning person, who several times comes up out of the water, but who, at last, is overwhelmed. Instead of endeavoring to save this art in its details, it is ignored by self-interest. But such an aversion to the best modern productions is sometimes punished very severely, as want of knowledge often brings common productions into the possession of individuals.

Since the death of Tariso, the great collection of violins, etc., which he gathered from all the regions of Europe, has been scattered again over all countries. Vuilliaume, who bought many of them, afterward resold some to violin makers and dealers; those instruments which were put in order by them are easily recognized.

This collection consisted mostly of all characters of Italian instruments, from the most commonplace to the celebrated Stradivarius. In many an admirer an interest may have been awakened thereby to possess one of these instruments. But it must not be expected that all of those violins still possess their original parts. Had not such amateurs as Tariso—and they are not rare in Europe—bought those instruments of that time and kept them safely, which contributed to their longer preservation, they would, especially if they had been always used, be in a much worse condition.

George Hart, of London, is also such a gatherer of and dealer in instruments. John Hart, the father of George Hart, whose personal acquaintance I made at Vuilliaume's, in Paris—when I was engaged to make for him a set of Stradivarius heads, from that of violin up to that of contra-basso, which should serve as models—undertook to gather such old Italian violins for the purpose of selling them again to other persons. From that firm there came, in fact, some specimens of the celebrated Italian masters to America, and they are interesting and very well preserved. I have seen and admired them; they are in possession of an amateur at Hartford, Conn. Here they are preserved again for the coming generation.

Violin players look with envy upon such violins in the hands of amateurs, but it is fortunate that most of them have come into such hands, for violins of this kind are very delicate, and although those which are well kept produce a beautiful tone, most of them have not that power of tone which is necessary for concerts.

The solo player, however, believes he must produce the strong tone of a violin by force, which breaks the tone, and is not heard distinctly. In this manner such violins are tortured and ruined. When such well kept violins continue to be well preserved, they may be the same after a hundred years. Such relics will then, no doubt bring still higher prices from those who wish to possess a violin of that kind.

But it is strange that some amateurs put a particular value upon a violin which has been in the possession of a rich nobleman, as if it is more likely to be genuine in that case? What a foolish idea! Such whims are not entertained by connoisseurs. There are enough aristocrats who possess only a fiddle, especially in America, and who know nothing about the value of a violin; it is rarely that they have at home a violin which is worth over five or ten dollars. When many of them hear that thousands of dollars are paid for violins, they think that persons who pay these prices must be crazy. The reason of this is that most of them know no difference between a ten dollar fiddle and a violin which costs as many hundreds of dollars!

Amateurs who pay thousands of dollars for a violin are here in America just as isolated as that enthusiast who paid six hundred dollars for the first ticket of the first concert given by Jenny Lind in New York, and the other who paid ten dollars for his admittance in order to be able to see the six hundred dollar man.

Thus I believe to have unrolled a panorama which will assist in the dissemination of knowledge and truthful views, which have only been obtained by a long experience.

OF THE MANNER OF PLAYING—TREATMENT OF BRIDGES ETC.

It has often occurred to me that violin players of all kinds find fault when the strings are not arranged in the manner to which they are accustomed, and almost every one believes his method to be correct. This subject shall be discussed here, so that a clearer insight may be obtained and the correct method ascertained.

There are violin players who have a greatly arched bridge, and others a very flat one, on their instruments. The latter, therefore, more than the former, have the advantage of being able to play on all violins, because they are accustomed to a bridge which is flatter. These different methods mostly arise from the different arrangements of the violins upon which pupils learn to play.

Ole Bull was an exception to this rule; with him it was not chance; of all violin players he used the flattest bridge on his violin; but it was his principle. His music pieces required it, and in his method he became a master.

I. B. Poznanski played at one time on a violin with almost as flat a bridge as that on Ole Bull's instrument, and I believe it will not have been forgotten that he produced, as if by charm, a great tone from his instrument. This proves that a great tone can be gained on a flat bridge. Therefore it depends only on the skill with which the bow is handled. Many violin players, however, are of opinion that they must press the bow on the strings very much, in order to bring forth a strong tone on the violin; but the pressure of the bow is limited; for when it is too strong, the ear becomes disgusted with the tone, nay, a scraping and jarring tone is produced by too strong a pressure, because the G string touches the finger-board in this case, in consequence of which many violin players wish to have the finger-board very hollow. But it must not be believed that in such a manner the right tone is produced; on the contrary, the full tone, which lies ready in the violin, is very easy to be gained by the knowledge and skill of handling the bow.

The rule is, that the tone must be drawn forth by the bow, and it must not be forced forth by pressure. The bow must not be led oblique, but straight over the strings, so that the hair lies flat on them ; it also depends on the flexibility of the arm, that the bow may not touch the strings stiffly, but in an elastic manner. Those who attract attention to their elbows cannot expect that the bow and the violin alone will do their service.

The most perfect condition of a violin requires the instrument to be so arranged that it can be played easily ; therefore, I determine that the height of the strings must be three-sixteenths of an inch at the end of the fingerboard, and that the arch of the bridge must have the same measure, three-sixteenths of an inch, between its two extremes, for bridges more arched than this cause difficulties to the player, because the movement of the bow is too much abstracted when passing from the E string to the G string. In such a manner, David in Leipsic had the violins arranged for his pupils.

On such arched bridges the two middle strings lie too high from the top towards the G string and E string, and it is an acoustical mistake, because it produces an inequality of the character of tone.

Such knowledge should be taught to the pupils in conservatories of music ; but it is generally believed that when a violin player has been made a professor he is able to satisfy the requirements of his position in this regard.

For the benefit of the learner, however, I will enter more nearly upon the knowledge which is required, especially in a conservatory, and to the imparting of which the teacher should attend. First I will mention as an example the conservatory at Leipsic when it was under the management of Director David. Most of his scholars were then compelled to play on new violins made by Bausch, which for their stiff and tough tone are for the greater part unfit for those who would become artists. This quality of tone, together with the fact that students were forced into a certain position and fatigued, caused them to become nervous ; but many parents who had no knowledge of it, sent their sons to that institute, even from America, and they had no idea that many of them brought back a nervous disease and were thus ruined. I heard this of no other conservatory in Europe. Thus it would appear that David pursued his own interest rather than that he cared for the good of his pupils.

Here in America we have violin teachers whose methods are preferable by far to such.

The following is a method according to which students should be instructed: The student must not be forced into a position of holding the violin so as to cause the ruin of health, but on the contrary, by means of a free position and natural holding of the violin the chest will be enlarged. This does not only benefit the health, but also facilitates the learning and progress.

It is of the greatest importance that students learn on violins which have good tone, for instruments which have a bad quality of tone usually discourage the beginner, so that he becomes nervous and soon considers playing an unpleasant work, and gives it up without knowing the reason why. Teachers, therefore, should have the necessary knowledge of the qualities which a violin must possess. A knowledge indispensable for them and a great benefit for the learner. For only a good tone has a charming influence upon the mind, and owing to this many beginners advance early to a high degree of perfection; therefore it must also be in the interest of the students to get familiar with the good tone of a violin, that their ear may not be accustomed to a sickly tone. Alas! This point is mostly disregarded by their parents, who have little or no knowledge of a violin, and it provokes some indignation in scientifically instructed teachers to teach their pupils on miserable fiddles.

If a teacher knows how a violin should be arranged, it is above all his duty to examine the instrument, and ascertain whether it can be used for the instruction of a learner; for as the violin is first arranged for him so he will ever be accustomed to have it afterward. For instance, on the violin of the solo player Ed. Mollenhauer, the strings lie on the fingerboard lower than on any other that I ever saw. No doubt he has learned on such an instrument. It is true that the virtuosoship is facilitated, but the strength of tone is impaired by such an arrangement.

The ingenius artist Brume, however, was so great a master that he played even on violins the strings of which lay very high, although he did not know this. Many, again, are accustomed to bridges that are very much curved towards the E string, because they did not know, when learning, how badly their violins were arranged.

A correct system must be the foundation of everything, but as the thories in this art are still dead letters for most violin players, there have arisen fantastical ideas, especially among the greatest of them. Ole Bull did his best to impart such ideas to others, yet many of them were, no doubt, excellent. Ole Bull always had a vehement desire to find something better beyond all possibility. Many of his ideas were contradictory to all the rules, and although he put some in practice he did not persevere in any of them for a long time, for a new idea occurring to him all others were supplanted by it.

It happened once that Ole Bull was visited in New York by another artist, who was called the "American Sivori." He, as well as many others thought that Ole Bull had a perfect knowledge of the structure of violins. Sivori, seeing that Ole Bull had a bridge on his violin which stood quite oblique—for the upper part of the bridge was bent backwards by a quarter of an inch,—adopted this idea. When his violin had been provided with such a bridge he came to me, and with great satisfaction he showed me this queer position of the bridge on his violin. I was highly astonished at him that he could approve of an idea which is against all correct theory and is nothing but a farce. I then explained to him not only the consequences which must arise from it, but also the impossibility, by such an arrangement, of bringing to bear an even horizontal pressure on the bridge. But he thought that which came from Ole Bull was better than that which came from my knowledge. Let us see what happened later. In a concert of his, while he was playing with enthusiasm, the bridge fell and broke!

Another day an Italian artist came with his Maggini violin to show me where the sounding post must stand in his violin, having obtained his information about it from Ole Bull. I could not help smiling when I saw that the sounding post was placed quite near the f hole. Upon expressing my surprise, he replied with the following insult: "What do you know about the position of the sounding post? You are no violin player like Ole Bull, therefore you cannot know about it." My answer simply was: "Only a fool can talk to me in that way, and very soon you will find out that you will have to give up such an insane idea!"

It was on the third day after that he came

back begging me to place the sounding post in his violin according to my judgment. When he had apologized for his indiscretion, I fulfilled his wish.

Thus I have become acquainted with several artists who constantly tortured their violins by getting the sounding post and bass-bar displaced. This proves a want of correct theoretical knowledge, and through this ignorance they make the sounding post wander about the whole violin.

The place of the sounding post can only be ascertained through the theoretical knowledge of the construction of the bottom and top of the violin. Many players think they can obtain the right tone by the position of the sounding post alone, but no sounding post can make good a fault in the construction of the bottom and top.

CIRCULAR WHICH ACCOMPANIED MY "EMPEROR VIOLIN" IN THE VIENNA EXHIBITION OF 1873—AN INTERESTING EXPLANATION ABOUT VIOLINS AND OF THE SCIENCE OF TONE.

It is an indisputable fact, that of all productions of art in the world, the violin has been least understood.

This wonderful instrument has remained an enigma to the musical world until now. How fortunate it is that this instrument does not understand human language, by which circumstance it escapes that medley of critical remarks which are made in its regard.

It is, therefore, in the interest of art and its votaries that I have determined to present herewith to the public the results of my long experience obtained in making violins, and in examining those sciences connected with it.

It is generally known that up to the earlier part of the eighteenth century the Italian masters made the best violins, and with the death of those artists a decline of that art, too, took place. Those so-called classical instruments have been, especially of late years, eagerly sought at high prices, by all artists and amateurs, because a settled opinion has taken hold of their minds that nobody is able to construct a violin which is fit for solo performances; that the secret which the old Italian masters possessed is not yet found, and that new violins, although constructed according to the rules of acoustics, cannot gain the desired perfection until after the use of a hundred years. This, therefore, animated many violin makers with an endeavor to overcome that difficulty, but in vain; at last Vuillaume, of Paris, was impressed with the thought of making wood look old by a chemical process, and he succeeded in creating a furor with his instruments made of such wood, so that people began to believe the right course was being pursued. It turned out, however, that after a few years those instruments deteriorated, and finally became useless and proved a failure.

This especially prejudices the minds of the virtuosi so far that they do not believe it to be possible to make violins which answer the general requirements of concert playing until they have attained a great age.

Vuillaume has, therefore, by his chemical

preparation of wood, injured this art seriously, because the previous prejudice was corroborated thereby. Such prejudices stand in the way of progress in making good violins.

But as everything in the world is going on, so the art of the construction of violins has not remained behindhand, and I can prove this to the musical world by my own experience.

To the knowledge of making such violins as artists and amateurs demand, there belong besides ingenuity in carrying out the mechanical work a knowledge of the following three sciences, namely: mathematics, acoustics and the choice of wood.

A knowledge of acoustics, which is most indispensable to the violin maker, cannot always be acquired, since it emanates from an innate genius, which makes itself manifest in the very choice of the wood.

When by the aid of these sciences I had arrived by a natural proceeding at what I aspired, I made violins in imitation of the old Italian instruments and presented them to great artists and connoisseurs, and the highest authorities of Europe and America. They pronounced them to be genuine old Italian violins, not only on account of tone, but also in regard to form and appearance. In this manner I broke that prejudice. I proved to the so-called "connoisseurs" that those violins laid before and acknowledged by them to be good, were of my making, hence they were new. If I had presented those violins as new productions of my own to those gentlemen, they would have condemned them forthwith and said that they would not prove good till they had reached a great age, and that they would perhaps in a hundred years equal the old Italian instruments.

In general, however, it is not taken into consideration that if a violin is not scientifically constructed the good quality of tone will never be obtained, either by much playing or by age. In applying the three above mention sciences I have gained not only the fine quality of tone, but also that ease with which the tones are made to come forth.

But we must be thankful to the great masters; they have laid for us the foundation of the manufacture of violins, by which they became immortalized.

Their system, however, is but little understood by the present violin makers, because very few intelligent people devote themselves to this art, and the most of those who are learning it, practice it not in the way of art, but of business. What wonder, when even the greatest artist in Europe, Vuillaume, imitated the very mistakes which the great Italian masters made in regard to mathematical division. He did not consider that they, in improving the art, made experiments in regard to form, swell and different thicknesses in working out the bottom and top. But there are a great many professional men who, from exaggerated veneration, consider all productions of those masters as law and beyond correction.

I have discovered that the old masters did not arrive at perfection, but made mistakes in their mathematical division and in the workmanship of the different thicknesses of the bottom and top. Those faults I have endeavored to avoid in the manufacture of my violins, and I think I have solved this problem.

Just so it is with the knowledge of tone. It is a great mistake to believe that it is only the player who has this knowledge. Experience has taught that playing and knowledge of tone are two different provinces, because the artist very seldom has an opportunity to make close study of the different qualities of tone, and is usually preposessed with his own instrument.

If many solo performers had more knowledge of tone they would not so often play in concerts on feeble instruments, which are too old, too defective in construction, or have been spoiled by bungling workmen who were employed to repair them. Such instruments often injure the solo performer exceedingly, and the critic is right in charging the to fault feebleness of tone. But the artist is generally satisfied if he only possesses an Italian violin.

Also in the science of tone I have found the way to gain that experience by which I have been enabled to make a violin which will satisfy an unprejudiced solo performer of the present and future.

I have confined myself to the natural process which the Italian violins underwent, and I have put the problem to myself that it must lie within the bounds of possibility to construct violins which will bring forth good tones at once and not depend on a promising future for all their good qualities, and I have not been mistaken, but have secured what I sought.

Many are still of opinion that the art of making violins and predetermining the qualities of tone, is a mere accident. This is, if taken in a general sense, true, because most of those who make violins scarcely know any more of it than a joiner, but the ability to construct violins according to the rules of art, requires a man who has enjoyed a technical education, and whoever has acquired the necessary capabilities knows the method by which the different qualities of tone may be produced and obtained.

Above all, he who occupies himself with repairs can least dispense with these capabilities, since he is often intrusted with the most valuable instruments; but alas! with what inconsideration do those who possess such instruments often give them, for repair, to botchers and fiddle makers.

This proves how great in this regard is the lack of correct judgment. Through such spoilers of violins most Italian violins have come to naught, because many who own such instruments think that whenever any one makes a neat piece of work and knows how to use his chisel, file and sandpaper, he is the man to be intrusted with such instruments. But where there is a lack of science, the repairer's work, be it ever so neat, may cause damage in half an hour which will be greater than can ever be made good again.

If a violin maker constructs bad instruments it is his own damage, but to make bad repairs is to ruin the instruments of others, the creations of masters.

Neither is a violin maker who does not know how to construct excellent instruments a good repairer. Yet there are many who think that good repairers need not possess the knowledge of making good violins. But what a mistake! It seems, however, wisely ordained by nature that even he who is less gifted and less learned may enjoy life, and thus gladly bear sacrifices in consequence of his error.

This is the plain and simple explanation of matters in regard to the manufacture of violins and the knowledge of tone, and those to whom this does not seem comprehensible may submit to a more thorough experience than they have gained until now; in this case they will, after they have fully convinced themselves of it, sometimes remember G. G.

A REPLY TO MR. E. SCHELLE'S CRITIQUE CONCERNING THE VIOLINS IN THE EXHIBITION OF VIENNA IN THE LEIPSIG "NEUE ZEITSCHRIFT FUR MUSICK, No. 52, 1873.

In the foregoing circular, treating upon violins, I said: "It is indisputable that no production of art in the world has been less understood than the violin." This truth has proved good again in Mr. Schelle's critique concerning violins, and it shows how little he is able to judge about them! In his very introduction it is plainly shown that he has made no studies in regard to tone when he says: "Thus an idea came to Vuillaume to make, by a chemical preparation, wood to look like that of the old violins. Instruments made of this material excel in regard to their splendid and real Italian tone."

Against this I assert just the contrary and can prove it to be nonsense by the fact that wood, when submitted to a chemical process, will produce a dry, covered tone, and the noble quality of tone—that which affects the heart—is lost.

Mr. Schelle then says: "We may also discover a similar experiment in the instrument which Mr. George Gemünder, of New York, has in the exhibition, under the ostentatious name of Kaiser Violin (Emperor Violin). Of course its manufacturer would protest against this insinuation, for in a little pamphlet he declares that by the assistance of three sciences, the mathematics, acoustics and knowledge of the wood to be chosen, he had not only comprehended the system of Italian school, but had even discovered errors in it,' etc."

Mr. Schelle further says: "There have been many celebrated violin makers who were gifted with the same talents and learned in the same sciences, yet they could not reach what they aimed at, in spite of their most strenuous efforts. We confess quite openly that in spite of his assurance we harbor the suspicion that Mr. Gemünder has taken refuge in a chemical preparation of the wood. The violin in question, a faithful imitation according to Guiseppe Guarnerius, is indeed beautiful in its appearance and has a very excellent tone. But the extravagant, really American, price of ten thousand dollars could only be excused when its excellence should have been proven good in future," etc.

From this (Mr. Schelle's) critique it is evident that he has tried to throw into the shade the interesting production of art which I had in the exhibition, in order to be enabled to put the productions of the Vienna violin makers in a more favorable light. But this proves that only such persons as are destitute of sufficient knowledge to judge of violins may be transported to such one-sided critiques, dictated either by partiality or other interests ; for if that were not the case Mr. Schelle ought to have blushed with shame in regard to that injustice and disrespect with which he illustrated the experience of an artist and spoke of his talents and sciences, to which Mr. Schelle is as much a stranger as he is to the artist's person !

As Mr. Schelle takes into consideration that the violin at ten thousand dollars exhibited by myself must first undergo "a proof of time," it may be rather advisable for Mr. Schelle to take a lesson of Gemünder, that he may learn those characters of tones which will prove good in future and which will not ; so that he may

be able hereafter to show better knowledge in
his critique upon violins!

From my childhood I have grown up in this
art in Germany and have devoted myself to
all those studies which are connected with it.
The last four years in Europe I passed at Vuillaume's in Paris, consequently I am acquainted
with the entire European knowledge of the
construction of violins.

Since 1847 I have made violins in America,
therefore my instruments do not require to be
subjected to a " proof of time," for it is without such a one that I have solved the problem
and secured at once the fine tone which all the
preceding violin makers strove in vain to find.
I obtained my purpose in quite a natural way.
This knowledge, however, does not lie in an
object whose secret is only to be secured by a
patent; it lies purely in the gifts of man. Another century may pass by before this problem
will be solved again. The closing page in Mr.
Schelle's critique sounds like a lawyer's plead-
ing in favor of a criminal. In this regard his
writing is quite creditable, for he has well
pleaded the cause of the violin makers of Vienna!

But then those words in my circular about

violin makers proved true again : "This wonderful instrument has still remained an enigma to the musical world until now. How fortunate it is that it does not understand human language, by which circumstance it escapes the medley of opinions which have been given in regard to it."

When, however, its clear tone was heard, and the easiness with which the tones came was noticed, then it became an enigma to professional men and they declared that this violin was an original fixed up again !

But later, when it was objected to and found to be a new Gemünder violin, it was ignored even in the newspapers. The *Neue Wiener Tageblatt*, of Vienna, called it afterwards " the false Cremona violin !" How envy here glared forth again ; for this violin was not exhibited as a Cremona violin, although it has been demonstrated that it had been previously really taken for a genuine Italian instrument.

Its introduction as " Emperor Violin " had a force and pungency which tickled the professionals, and what surpasses all belief is, that they themselves crowned the work. It was, indeed, the greatest premium that I could gain, in spite of all the pains which those men gave to themselves to deprive me of my merit. Thus a moral prize values higher than a piece of metal?

Although many mocked at the high price, yet no such violin could be made by all those deriders, should millions of dollars be offered to them. Therefore an unrivaled artist has the right to fix any price on his productions. Although an offer of $3,000 was made for it, yet nobody was charged to sell it, even if $10,000 had been presented.

The newspaper of the exhibition of Vienna, published on the 17th of August, 1873: "Gemünder found fault with the Italian constructions and those of Vuillaume."

If Gemünder had not extended his studies so far he would probably not have stirred up those matters which had given such a headache to those people of Vienna, for George Gemünder became thoroughly acquainted with both the faultless and the faulty points of the Italians in the construction of violins. If those people of Vienna had had the good luck to discover imperfections on the above mentioned constructions, then they would have made a great cry about it.

The same newspaper says in another passage : " The tone of this violin is indeed strong and beautiful and has an easiness that pleases, also it has not that young tone peculiar to the very best new violins." In saying these words the writer confesses the truth in his innocence, and this verdict crowns this violin again, because this character of tone is just that one which all violin makers in the nineteenth century have been trying in vain to find.

And further : " For this reason some professional men gave vent to the suspicion that the wood was submitted to an artificial preparation, probably by the use of borax." Such was the nonsense to which this peerless violin was subjected, since there was none to take up its defence. *The annexed description in which all chemical preparations were peremptorily opposed, was entirely disregarded by them.* Thus there is no other way to advise those pseudo-professional men to have such borax violins made and patented !

To those gentlemen who call themselves professional men, I, George Gemünder, declare that I am ready at any time to sacrifice my " Emperor violin " or any other which I have made, and I propose to give it to the best chemists in the world to be cut to pieces, that they may examine the wood and ascertain if any chemical preparation has been used. If this is found to be the case they may be allowed to scold and blame me publicly as much as they please; but, if nothing of that kind is found, they are to pay ten thousand dollars for the "Emperor violin."

    Address: GEORGE GEMUNDER.
                          ASTORIA, NEW YORK.

www.ingramcontent.com/pod-product-compliance
Lightning Source LLC
Chambersburg PA
CBHW020150170426
43199CB00010B/967